PAST & PRESENT

GLENDORA

Opposite: By 1889, the pepper trees planted up and down the 14 original streets laid out by Glendora's founder, George Whitcomb, were taking root and climbing skyward. Looking northeast from around the corner of Leadora and Vista Bonita Avenues is the sparsely settled town, merely five years old. The boy is Archie Powell, and he is holding a Model 1873 lever-action Winchester, known as the "Gun that Won the West." He sits atop Dan, a burro owned by Keith Suydam, whose land is in the foreground. The building over the center of Archie's rifle is William Cullen's barn, while Cullen's house sits just above the rifle's muzzle. (Courtesy of the Glendora Historical Society.)

PAST & PRESENT

GLENDORA

Ryan Lee Price

This book is dedicated to the residents of Glendora who endeavor to maintain and preserve the rich history of Glendora and to those at the Glendora Historical Society who strive to teach each new generation about Glendora's history and share with them the events, places, and stories that make Glendora special.

ISBN 978-1-4671-0665-8

Library of Congress Control Number: 2020949761

Published by Arcadia Publishing
Charleston, South Carolina

Printed in the United States of America

For all general information, please contact Arcadia Publishing:
Telephone 843-853-2070
Fax 843-853-0044
E-mail sales@arcadiapublishing.com
For customer service and orders:
Toll-Free 1-888-313-2665

Visit us on the Internet at www.arcadiapublishing.com

On the Front Cover: The electrically lighted sign over Foothill Boulevard (at the corner of then Michigan Avenue) welcomed visitors and Glendora residents alike from the time it was erected in 1924 to the time it was taken down around 1935. Smaller metal versions of Glendora's "Pride of the Foothills" sign were sold during this time as ornaments that residents could attach to their car's license plate frame to show their hometown pride. Though the sign is long gone, hometown pride among Glendorans is still as strong as ever. (Past, courtesy of the Glendora Historical Society; present, author's collection.)

On the Back Cover: In 1887, Glendora founder George Whitcomb named the 14 streets in his new town after places he had lived or that were memorable and after people dear in his life. Leadora, Carroll, Ada, and Virginia were all members of his family, while Vermont, Pennsylvania, and Michigan were places he had lived or worked. The anomaly is Minnehaha, which was renamed Foothill Boulevard around 1911. It was likely named after Minnehaha Falls in St. Paul, Minnesota, a beautiful location near where he met his wife, Leadora Bennett, in 1856. (Courtesy of the Glendora Historical Society.)

Contents

Acknowledgments

A grateful thank-you is in order to those who made sure what I wanted to impart in the chapters to follow is correct, accurate, and proper, particularly my friends and fellow history lovers at the Glendora Historical Society but specifically Stephen Flowers, Sylvia Barton, Gordon and Norma Rowley, and David Fredendall.

I would especially like to thank my wife, Kara, for her patience during my frequent trips to the more obscure points of Glendora, and my children Natalie and Matthew for their curiosity and encouragement. A book like this, with so many moving parts, couldn't be possible without the editing skills of my mother, Linda, and I appreciate the care and support she always gives me and my projects. But most importantly, I would like to extend my deepest gratitude to my brother Jason for accompanying me on my circuitous wanderings around town, for making sure I didn't get hit by a car while taking photographs, and for being a constant companion during those long shooting days.

All historic images used in this book are courtesy of the Glendora Historical Society. All present photographs are from the author.

These two adorable children are the author (right) and his brother Jason standing in front of the Glendora sign on then Alosta Avenue in 1977, right after their family moved here from San Gabriel. It may not sound like a long time ago, but in those 40-odd years between then and now, a lot has changed.

Introduction

Years before it would be known as "the Pride of the Foothills," *Glendora Gleaner* editor John Atlas Jones described Glendora as "the Crescent City" in the April 1912 issue of *Out West* magazine. He wrote about Glendora's recent incorporation: "With the organization of a city government, Glendora may fairly be said to have entered upon a period of growth and prosperity seldom witnessed, and it is fair to presume that in the next three years the population will not be less than 5,000 and the wealth increased many fold."

The City of Glendora was incorporated merely six months earlier, officially recorded with the County of Los Angeles on November 13, 1911, as the 25th city to achieve incorporation in the county (the number currently stands at 88). It was not an easy road for those wishing for complete independence, as conflict and divergence quickly divided the town. Eleven years after George Whitcomb laid the foundations for Glendora, it surpassed village status and was incorporated as a town in 1898, followed by an election for the town board. Though a small victory for progress, Glendora was one step closer toward self-government and independence from Los Angeles County. Under the control of the county board of supervisors (at the time of incorporation Charles Dewey Manning was Glendora's district supervisor), Glendora was offered merely the basics in government support, such as regional fire and limited sheriff protection, judicial and financial administration, and intercity road maintenance.

Being on the eastern outskirts of the county, members of Glendora's Board of Trade (a precursor to the chamber of commerce that was founded in 1903) and other prominent Glendora residents felt that they needed more control over the future of their town. Since there were very few problems that could not be solved by the Glendora Land Company or a collection of like-minded individuals, why not incorporate the town into a city and control their own fates? Mostly thanks to their urging, the telephone came in 1902, gas followed in 1904, and electricity in 1908. Meanwhile, an influx of new citizens poured from the newly completed Pacific Electric Railway, eager to join the largest resurgence of the real estate market since the 1880s. The town of Glendora was quickly coming of age as new lots were sold, old ranches divided, and houses built. The temporary feel of wooden stores and frame buildings gave way to the permanency of cement and brick. The Weaver Building (1903), the Converse Building (1905), the First National Bank Building (1907), and the Gem Building (1909) provided new residents of Glendora a sense of enduring commitment to Whitcomb's 14 dusty little streets.

In the August 10, 1911, edition of the *Glendora Gleaner*, editor J.A. Jones published the official boundary lines in a "Notice and Petition for the Incorporation of the Town of Glendora into a City of the Sixth Class." Signed at the bottom of the petition were 184 names, representing many of Glendora's prominent citizens. The boundary lines, as established from the first series of meetings, included roughly 1,900 acres within the city limits, tracing a conspicuously meandering line from the corner of Loraine Avenue and Palm Drive, west to Grand Avenue and Charles Silent's front gate, south on Grand to Sierra Madre Avenue, west on Sierra Madre to Ben Lomond Avenue, south on Ben Lomond to Foothill Boulevard, east on Foothill to Grand Avenue, south on Grand to Mauna Loa Avenue, east on Mauna Loa to Loraine Avenue, and then north on Loraine to the point of origin.

On October 2, 1911, after meetings on the two previous Mondays were held where vigorous points were made by individuals about why their property should or should not be included within city limits, the Los Angeles County Board of Supervisors granted the petition for Glendora to vote for incorporation. As well, they settled the boundary dispute once and for all by revising the proposed city's borders, as reported in the September 6 edition of the *Gleaner*: "Sierra Madre avenue on the north; Grand avenue on the west; Mauna Loa on the southeast to a point where it reaches the San Jose hills; thence with a deflection for a short distance to the intersection of Loraine avenue; thence on Loraine avenue to Sierra Madre avenue, or the point of beginning."

The approved petition encompassed 1,440 acres of city limits, to the delight of some and the disappointment of others.

The following edition of the *Gleaner* ran an official description of the city limits as well as the official call to vote. The notice, announcing that the election was scheduled for October 31, urged residents to vote "For Incorporation" or "Against Incorporation," as well as to fill the six board of trustees positions, a clerk and a treasurer being required by law for a city with a sixth-class incorporation status.

In large print in the same issue was a message from Charles Manning, a supervisor representing the first district of Los Angeles County: "Having thoroughly investigated and gone over the territory sought to be included within the boundary lines of the proposed City of Glendora, we decided upon the lines established believing that we had done . . . justice to all interested as we saw it, and can see no reason now why we should reverse our decision by extending the lines in any direction."

"Grand Victory For Progress" rang the headline in the *Gleaner* on November 3, 1911. What was called the "most hotly contested campaigns known in the history of Glendora" and a "sweeping victory" was really quite a close call. The vote for incorporation was slimly won by only 42 votes, with 152 for and 109 against. Interestingly enough, 312 votes were cast, but only 261 were counted due to voter mistakes; a full 16 percent of the total ballots collected were thrown out, enough to swing the vote in the other direction.

In the same issue, Jones editorialized about the events of 1911: "We . . . bid adieu to the pretty . . . village we have loved so long and well—the apple of our eye for [the] past 16 years—confidently believing that when we returned to our offices and labors, it would be our good fortune to publish to the whole world that Glendora had dropped her swaddling clothes, let down her silken skirts and was now ready to assume her rightful place among the cities of this fairy land."

This loss of Glendora's original identity has not slowed, even today. While there are a handful of citrus trees scattered behind backyard fences, a few stalwart reminders of an industry that supported the city's lifestyle for so many years, the orange is no longer king. It was quietly replaced by houses, a sense of community, and a palpable swelling of pride felt by every citizen who refers to Glendora as their hometown. This translates into the very essence of what it means to be a Glendoran, and whether one's family has been here for five generations or five days, the Pride of the Foothills is not just a kitschy slogan used to sell buttons and bumper stickers; it is a definition of what Glendora has become.

CHAPTER 1

On the Streets of Glendora

It is difficult to imagine that there was ever a time Glendora looked so simple and so empty. This is an alternate version of a promotional map drawn by Edwin S. Moore in January 1888 showing the best features of the new town of Glendora. From the 6,000 little pepper trees dotting the dusty roads to some of Glendora's first buildings, churches, and homes, this map exudes an amiable quaintness Glendora has enjoyed for 134 years, perhaps even in spite of progress.

The construction of the 210 Freeway began in 1958, and the first section, known as State Route 118, opened in 1966. In 1971, the route came through Glendora and was completed to Foothill Boulevard in La Verne, where traffic dumped into the western stub of Route 30. Looking northeast from atop the South Hills from 1968 shows an empty swath of land that was cleared of houses to make way for the elevated freeway. Notice the original route of Glendora Avenue.

In 1929, the Santa Fe Railroad built the rail overpass over then Alosta Avenue. Shown is the famous Bar T Oak with Compromise Line Avenue behind it. The compromise was between landowners north of the street and the San Jose Ranch Company, which owned land south of the street. For decades since the Mexican land grant days, this gnarled oak represented that boundary and was repeatedly burned in an effort to remove the marker and thus change the property lines.

Charles Kreglo bought 40 acres of land north of then Highland Avenue around 1900 and called the ranch Kregmont. Before he could build a home, he moved back to Los Angeles and deeded the land to his daughter Katherine and her husband, Milton Martin. In 1932, Kregmont was sold to Arthur Bourne Jr., heir of Singer Manufacturing. This postcard, labeled "Palm Drive," is a view of Kregmont Street, which is still lined with palm trees.

A 1955 Ford glides down Grand Avenue at the Ada Avenue intersection. On the left is the house of Emerson Gard, built in 1910. Among other business interests, he was vice president of the First National Bank on Glendora Avenue. His son Frank J. Gard was the only Glendora resident killed in World War I. The Gard house still stands, while the orange groves on the right made way for the Emanate Health Foothill Presbyterian Hospital.

Originally referred to as the Old City Road, Foothill Boulevard in Glendora was called Minnehaha by Glendora founder George Whitcomb, probably named after Minnehaha Falls in St. Paul, Minnesota, near where he met his wife, Leadora Bennett, in October 1856. The street was paved by the county in 1911 and renamed Foothill, although there are some places where "Minnehaha" can still be seen stamped in the original concrete sidewalks. This picture looks west toward downtown at the intersection of Minnesota.

FOOTHILL BOULEVARD LOOKING WEST. GLENDORA, CALIFORNIA.

As one of the oldest panoramic views of Glendora, this bird's-eye view was taken in 1893 on the summit of the South Hills. Looking closely, one can spot the Santa Fe Depot and a locomotive headed east, the Glendora Hotel, the grammar school, and the Methodist church in its final location on Bennett Avenue. The town of Alosta was still booming, as Central, First, Alosta, and Chestnut Avenues are visible.

Hiking into the mountains above Glendora has always been popular, and there were many trails connecting Sierra Madre Avenue with the Glendora Ridge Road. Historic routes built in the late 1800s, like the Garcia, Whitcomb, Englehardt, and Woods Trails, were heavily traveled by hikers and burro riders. In Glendora, Donald Pflueger wrote, "In the spring, during wild flower season, and when the air was clear, it was an exhilarating experience to look out over the valley all the way to the blue Pacific."

Brand new electric streetlights can be seen for the first time in Glendora, shown here on Foothill Boulevard west of Wabash Avenue. On March 23, 1914, the Newbery-Bendheim Electric Company finished installing 210 streetlights on Michigan, Vermont, Vista Bonita, Bennett, and Minnehaha Avenues—laying seven miles of wires that connected to the master switch at city hall. At 5:00 p.m. that night, according to the March 27 edition of the *Glendora Gleaner*, "the globes resting securely at the top of the lamp posts were all aglow."

Though Glendora was well known for its citrus orchards, specifically its lemons and oranges, other crops were grown within the town's limits, as many of Glendora's residents were expert horticulturists. In the November 30, 1890, issue of the *Los Angeles Herald*, George Whitcomb boasted about his acreage of red-skin tomatoes, while Lawson LaFetra claimed to have planted five acres of peas. Shown here is a field of strawberries on John Hommell's property in the town of Alosta, with the South Hills in the distance.

Sometime before 1887, George Whitcomb planted about 6,000 pepper trees (Schinus molle) along the streets that would become Glendora. They were fast-growing trees, giving prospective land buyers the illusion of prosperity. Approximately 15 years later, as this photograph shows, they were flourishing on then Michigan Avenue at Leadora Avenue. In 1909, Joaquin Miller, the poet of the Sierras, said during a visit that Glendora is "a poem in pepper trees." By 1930, most of these trees were gone.

Maybe Minnehaha sounded quainter, but by the time this postcard was printed in 1915, the street was paved and renamed Foothill Boulevard. Nearly directly across the street from these delightful Craftsman-style bungalows and kit houses stood a small concrete obelisk with the number 27 stamped into it. This let drivers know that it was exactly 27 miles to Los Angeles. In the distance is the Chance Building on what was then Michigan Avenue.

On the right is an orange stand just outside city limits in 1934 on the northeast corner of Foothill Boulevard and Loraine Avenue. This is part of a series of accident claim photographs taken by the prolific Dick Whittington Studio, the largest photography studio in Southern California from 1924 to 1987. His 1932 Ford sits on the side of the road, and the sign points to the ranger station, Dalton Canyons, and Camp Bonita, 16 miles away.

When the Santa Fe Railroad built the rail bridge across Alosta Avenue in 1929, its Y intersection with Valley Center and Compromise Line Avenues was reconfigured. Alosta was straightened and extended eastward to Lone Hill Avenue and beyond, bypassing Valley Center, which now terminates at Compromise Line instead. The remainder of Compromise Line to Lone Hill became Alosta, which was renamed Route 66 in 2003.

CHAPTER 2

DOWNTOWN, THE HUB OF THE CITY

Around 1913, downtown Glendora was still a rural oasis in a sea of citrus groves, but would not remain so for long. By the 1960s, almost all of the city's groves packed with citrus trees gave way to progress, development, and the temptation of the rising value of land in today's Southern California. In his popular column "On the Move" in the *Los Angeles Times*, Ed Ainsworth wrote: "If you want to live in an orange grove in Glendora . . . you'd better hurry."

The intersection of then Michigan Avenue (now Glendora Avenue) and Foothill Boulevard was controlled by a stop sign in 1935 and lined with queen palms instead of the ficus trees standing today. Venberg's menswear faced Turner's Café in the Chance Building, while the Christian church's steeple rises in the background. On December 17, 1990, a fire nearly burned the Chance Building to the ground. After extensive renovation and earthquake retrofitting, it contains a variety of businesses.

As early as 1903, the first cars were appearing in Glendora, but by around 1915, when this picture was taken, the prosperity of early Glendora was evident by the number of automobiles on its roads. Looking at the east side of then Michigan Avenue from Meda Avenue, many of these early structures—the Banholt Building, the Pacific Electric station, Jones Livery, and the Glendora Livery Stable—did not survive. All that is left in this image is the Chance Building on Foothill Boulevard.

Glendora Avenue of the 1960s was really the birth of the current look of the street; all of the existing buildings had been built or were scheduled to be constructed, and the last of the old wooden structures had been swept away. The palm trees that replaced the pepper trees had given way to the ficus trees, and Michigan Avenue was renamed Glendora Avenue in March 1964. The businesses may have changed, but the ficus trees are standing tall.

The flagship of Glendora Avenue has always been the First National Bank building, originally the home of the First Savings Bank of Glendora, chartered in 1908. Shown here around 1909, the ground floor houses the bank and Hall's Drugstore, while upstairs was the Glendora Opera House, intended, according to the *Glendora Gleaner*, for "public meetings, concerts, theatrical troupes, and secret societies." It was the first business to receive electricity, on April 9, 1908.

Though rarely shown from this angle, this is Michigan Avenue looking northeast from in front of Jones Livery, showing various 1915-era shops, all protected from the mid-day sun by what looks like the same style of awnings. From a millinery and a men's furnishings store to the stately Converse Building—housing the Fountain Confectionery—the First National Bank building, the Gem, the Weaver Building, and Glendora Bank at the north end of the street can be seen.

On Michigan Avenue, the streetlights—erected in 1914—were spaced 120 feet apart, as opposed to the standard 200 feet, to provide more light in the business district. In front of the Anderson Pharmacy in the Banholt Building sits an interesting juxtaposition of the old and the new: a 1916 Buick in front of an old-fashioned horse and buggy. The Banholt Building—the first building to have sidewalks in town—met the bulldozers around 1965.

Johnny Bock came to California in 1918 and worked at many different stores until he purchased from Ernest Cobb a variety store originally at 153 North Michigan Avenue, calling it Bock's. When he needed to expand, Bock moved the store across the street to this location, where he was in business until 1979 when he sold to his manager Loren Brown. Bock's went out of business in the mid-1990s. The site has hosted a variety of businesses, including Classic Coffee.

The original business district of Glendora was on Vista Bonita Avenue above Bennett Avenue. This is where the Englehardt brothers had their store, among others. By 1900, fires and lack of foot traffic persuaded businessmen to shift operations to Michigan Avenue. Shown looking north, just south of Meda Avenue, several structures remain today, such as the Weaver Building and the Kamphefner Building on the left and Seimears Furniture on the right.

Looking south on Michigan Avenue from Bennett Avenue around 1910, a prosperous rural downtown exists. Glover's barbershop stands at left between the two-story Kamphefner Building (not shown) and the Seimears Building. McNair's Glendora Café was housed in the original Glendora Land Office on the corner of Meda and Michigan Avenues, while the Women's Christian Temperance Union erected the drinking fountain at right. The pepper trees obscured the buildings then as much as the ficus trees do today.

The Weaver Building was constructed around 1903 by realtor Charles A. Weaver (at right below); soon thereafter, partners Charles M. Matthews and Clyde Wamsley opened a grocery store here on the corner of Meda and Michigan Avenues. In the foreground of this December 1909 photograph is a 1907 Ford Model N, and posted on the wall are various advertisements, including one for a stenographer for Rolfe Bidwell. For many years, this was Reed's Hardware, but now it is Frisella's Roastery.

Then Michigan Avenue, this is Glendora Avenue at Christmastime around 1960. Of interest in this picture is the Chance Building on the right, constructed in 1912 by Frank LeRoy Chance, a Chicago White Sox player known as part of the "Tinker to Evers to Chance" triple play that made him a star. Chance settled in Glendora on his Cub Ranch orange grove on Grand Avenue, just north of Baseline. At left is the Arthur Venberg Building, erected in 1925.

By 1980, about when this picture was taken, the population of Glendora was 38,654, and the town had annexed enough new property to occupy 15 square miles. Recreation Park, just east of here, was renamed Finkbiner Park on September 7, 1980, and the popular *Glendora Press*, whose offices were on the west side of the street, was sold to Highlander Publications in October. Businesses came and went, but the ficus trees kept growing skyward.

Constance Talmadge's *The Goldfish* opened in 1924 and is playing at the Mission Theater, a 750-seat theater built with the Finance Building in 1923 on the southwest corner of Foothill Boulevard and Michigan Avenue. Also, there was the Glendora Bank, Ray's Candy Shop, and James D. Dyer Grocery. In 1930, Richard and Maurice McDonald purchased the theater and renamed it the Beacon Theater. It was torn down in 1968 to become a parking lot for the strip mall there today.

On the right of this long, sweeping shot of Michigan Avenue around 1917 is the first bank in Glendora and the first brick building, constructed in 1902. The bank was organized by Frank Eugene Odell, and was successful for a brief time. Like so many, the bank failed during the Depression, and the building was occupied by a real estate office, the *Glendora Press*, and the Glendora Water Company before being razed in 1960.

This photograph was taken within minutes of the one on page 38. Spanning Foothill Boulevard was this impressive sign that was illuminated at night. The Financial Building and James D. Dyer's grocery store flank Foothill Boulevard in 1924. Farther down Foothill is a Chevrolet dealer, which in 10 years would be Tucker's Glendora Machine Shop next to Wilbur H. Glines's Ford dealership (which became Biszantz Ford in the 1950s). The sign was removed around 1935.

Orange Tree Bazaar was a series of antique shops carved out of the original Wood Building that was built around 1910. Abutting its north side was the corrugated metal F.A. Seimears furniture store (and undertaking and cremation), which had been at that location since around 1900. The Wood Building eventually enveloped the Seimears building to form the Orange Tree Bazaar, but around 2010, it was completely remodeled into a variety of businesses.

Gold Rush prospector turned teacher turned lawyer Charles Converse brought his family to Glendora around 1900 and was successful enough to commission the Converse Building on Michigan Avenue. With his offices upstairs, Converse was able to rent out the rest of the building to a variety of businesses. On May 4, 1912, Converse was struck and killed by a train while crossing Loraine Avenue. The building today is hardly recognizable, as it was completely renovated in 1965.

Wooden boards span the dirt gutters that line the upper business district of the wide Michigan Avenue around 1903. To the right, south of W.P. Nye's real estate office, is the two-story Glendora Bank building that had just opened. To the left is Glover's barbershop and south of that is an awning on the front of either a livery stable or the delivery wagon for Seimear's furniture business. In a few short years, Michigan Avenue would become the center of Glendora's civics and commerce.

This is Michigan Avenue around 1915. A relic of a bygone era is the Glendora Land Company office, built by George Whitcomb in 1887 for the first sales of Glendora's lots. Originally, it stood at Vista Bonita and Leadora Avenues when Vista Bonita was Glendora's business district, but it was moved to Meda and Michigan Avenues sometime around 1905 and then to Whitcomb and Minnesota Avenues before 1920, where it stands today as a private residence (minus the turret).

CHAPTER 3

Rain, Storms, Mud, and Floods

Being nestled so close to the foothills of the San Gabriel Mountains and the mouth of Dalton Canyon and many smaller arroyos, Glendora is no stranger to flooding disasters. The first big event on record was a result of the double cyclone strikes in February and March 1938, when 14 inches of rain fell in five days. Floodwaters in Glendora, shown here at the corner of Foothill Boulevard and Michigan Avenue, completely flooded the basement of city hall. West of town, the surging San Gabriel River destroyed three piers of the massive Puente Largo Bridge.

Six months after the Canyon Inn fire burned 19,100 acres of dry grass above Glendora (killing eight people) in July and August 1968, a torrential storm dumped 31 inches of rain in February 1969 alone. Many of the streets near the foothills became rivers of mud. Shown here is Rainbow Drive just before it meets Bubbling Well Drive, where residents used wood planks and barricades to divert the flooding. On the left, the Rimer, Constenia, Herzog, and Miller homes were all damaged.

In the 1950s, county officials recognized the risk of flooding in Glendora and constructed nine large debris basins and flood channels to protect the city, but the severe flooding in 1969 caused all nine basins to quickly reach capacity and spill over. Many of Glendora's northern streets were overflowing and flooded, as was Foothill Boulevard, just east of the high school. An advertisement for the Oak Tree Rancho development is on the right.

The real estate office for the Oak Tree Rancho development sits to the left on newly minted Whispering Oaks Drive, where houses were waiting to be built, displacing a few dozen acres of orange groves. Perhaps thanks to the flooding that took place in the winter of 1969, many of these houses would not be built until 1970–1971. Though the ornamental lights have changed, the redbrick entry walls and finials remain.

The world-famous Rubel Castle property backs Calabria Drive, which was heavily damaged during the 1969 rain and subsequent mudslides. On the left is the Rubel Box Factory building. In addition to the onslaught of debris littering the street, a relatively new 1967 Pontiac GTO did not make it far out of the driveway during the floods (or it was pushed there). The Rubel Castle and Calabria Drive were eventually cleaned up, and the house on the left was remodeled around 2018.

Almost every north-south street in Glendora that met the base of the foothills experienced some damage during the mudslides of 1969. City engineer Karl Kohlhoff warned, "Homes in the paths of the canyons opening into Glendora are in serious peril and stand to be severely damaged." The Theiller house, for example, sits on the inside curve of Rainbow Drive, just south of Bubbling Well Drive, and took the wall of water and mud head-on.

CHAPTER 4

INTERESTING DESTINATIONS

Taking advantage of a prime location, Walt Wiley opened an orange juice stand where customers would do the usual: park their car and walk up to order food and drinks at the window. It proved to be such a popular concept that when Richard and Maurice McDonald were looking for ideas to make their own restaurant unique around 1940, they modeled McDonald's on Wiley's orange stand. This building later became the Old Hickory Inn.

Officially dedicated Big Tree Park on April 26, 1975, the tree shown here around 1970 is a Moreton Bay fig (Ficus macrophylla), a native tree of Australia planted by Edgar J. Owens in 1890. It is the second-largest tree of its kind in California, residing on the corner of Santa Fe and Colorado Avenues at what was the front yard of Arthur Bohnard, a talented watercolor artist. With help from arborists, the Big Tree is still as strong as ever.

On Monday, September 21, 1914, east of Loraine Avenue on Sierra Madre Avenue, the crew of the Layne & Bowler Pump Company, along with city officials, turned on the pump that would bring fresh well water to Glendora. By the 10th day of testing (on September 30), 87 miner's inches of water was flowing, yielding twice what the city needed at the time. Since then, water resources in Glendora have been a precious commodity at the Arthur E. Cook Jr. Water Production Facility.

At a cost of $6,000, the Woman's Clubhouse on Glendora Avenue was built in October 1914 as a simple bungalow with an assembly room, stage, tearoom, kitchen, office, and two dressing rooms. The Glendora Women's Club was organized in the spring of 1908 with 40 members, and at the time of the dedication of the new clubhouse, their ranks numbered 100. The first purchase for the clubhouse was a baby grand piano. The building was extensively remodeled around 1955.

Around 1937, Walt Wiley started an orange juice stand on the corner of Route 66 and Grand Avenue, one of the first order-at-the-window stands in the valley. In 1949, around the time this picture was taken, Homer and Opal Weimer started the Old Hickory Inn as a full-fledged restaurant. It stood for 60 years, though the name was later changed to the Derby East. Despite the efforts of local historical organizations, it was torn down to make way for a parking lot.

Elizabeth Shorey stands over the Doughtery, Ott, and Kamp tombstones in the Fairmount "Pioneer" Cemetery on San Felipe Hill, with the Monrovia Nursery in the background. Shorey was a longtime trustee of the Fairmount Cemetery and oversaw its protection from 1988 to 2001 during the time the nursery began to develop its property into houses. Today, her daughter and trustee Sylvia Shorey Barton continues to make sure the cemetery remains a place of honor in Glendora's history in its newly developed neighborhood.

The land on San Felipe Hill was donated by James C. Preston in 1886 to be a cemetery, even though burials had been taking place there for years. After 1888, it cost $10 to be buried, and there are well over 200 burials. In the distance still stands the markers of the Wright family plot, while many others have fallen prey to vandalism and theft. With renewed support, today the Fairmount Cemetery Trust is providing many upgrades and regular maintenance.

By the late 1950s, the shopping experience had drastically changed with the invention of the supermarket and the strip mall. The Mayflower shopping center on Grand Avenue and Gladstone Street opened around 1956 with a wide array of shops, featuring a cleaners, a toy store, and clothing stores in addition to the Mayflower grocery store and the Cornet five and dime. Today, besides Spaghetti Eddies, the Red Lantern, and a couple others, most of the Glendora Village Civic Center is empty.

CHAPTER 5

Churches, Civic Buildings, and Schools

Contractor J.C. Bannister and mason G.H. Whyte laid the cornerstone for the new city hall at Foothill Boulevard and Michigan Avenue in November 1920. It was built with fireproof bricks, with reinforced cement girders two feet thick, and finished with California stucco in a light tan. The ionic columns suggest colonial architecture, but the red roof tiles provide a Mediterranean look. It is 78 feet wide, 70 feet deep, and 46 feet tall, and cost $69,500, which included the furnishings and adjacent park.

The first bond the city issued was in April 1913 to raise $4,500 for a fire station and city hall. Designed by Glendora city engineer Floyd Dessery, the new firehouse was constructed using reinforced concrete. It was large enough to house two engines, though Glendora could only afford one at the time, a 1915 American-LaFrance Type 40, at $6,000. Today, city hall houses the Glendora Museum, and the fully restored fire engine regularly makes appearances at parades and city functions.

By July 1930, the library had outgrown its rooms in city hall, so C.C. Harding was contracted to build the east wing of city hall at a cost of $2,880 (through the generosity of city attorney Rolfe Bidwell). Shown here in the mid-1960s, the queen palms that lined Foothill Boulevard and Michigan Avenue have since been removed to make way for the ficus trees.

In 1961, Glendora celebrated its 50th-anniversary golden jubilee of incorporation with parades and festivities, including the burial of the 33-by-89-by-28-inch Centurion time vault on September 30, not to be opened until the diamond anniversary in 1986. Citizens were asked to donate photographs, documents, and memories descriptive of life in Glendora. Instead, it was opened in 1987 (the 100th anniversary of the town's founding), when it was discovered that most of the artifacts had been ruined from water seepage and mold.

The popularity of the automobile increased people's ability to travel, and before motels were plentiful, auto camps sprang up like this one north of Foothill Boulevard and west of Vista Bonita Avenue. A February 25, 1921, *Glendora Gleaner* headline claimed, "Autoists Delighted With Parking Grounds," but by 1926, vagrancy, noise, and a general unkempt appearance caused the city to close it. The Grace Episcopal Church in the background was moved to Mountain View, and all that remains is the old live oak tree.

The first movie theater in Glendora was operated by Jerry Ewers on Michigan Avenue north of the Little Dalton Wash. It was nothing more than a canvas tent. He soon moved the Glendora Theatre into the A.B. Rogers Building on the southeast corner of Foothill Boulevard and Michigan Avenue, where city hall would be built seven years later. Here, Ewers is showing *The Birth of a Nation*, which dates this picture to early 1915.

A spiritual difference caused a split in the Church of the Brethren congregation around 1930, and on October 1 of that year, the Independent Church of the Brethren was formed with approximately 140 charter members. The church was built on the northeast corner of Michigan and Whitcomb Avenues with rock from the San Gabriel Canyon to save money on construction costs. It is now the Cornerstone Bible Church.

The First Christian Church's charter members included many of the early Glendora settlers, such as Cullens, Englehardts, and Wests. Leadora Whitcomb donated the lot on the northeast corner of Wabash and Bennett Avenues for the church, completed in 1888. Eventually, land was purchased for $2,000 on the northeast corner of Michigan and Bennett Avenues for a larger church. On December 17, 1922, the cornerstone was laid, and the church building was dedicated on September 16, 1923.

On April 1, 1887, George Whitcomb had everything prepared: The hotel was open, a fountain proved the town had plenty of water, and the pepper trees provided a sense of permanency. The first lot sold for $475 to Jefferson Patton. It was on the northeast corner of Bennett and Michigan Avenues. By the end of the day, 291 lots were sold, ranging from $325 to $675 each. Here, in about 1952, Joyce McClendon stands near the rock that commemorates the event.

One of George Whitcomb's ideas for his town was that it have a proper school, so funding was set aside from the sale of the land to provide for a school building and a teacher. On December 1, 1887, the first class took their seats at the newly completed Glendora Grammar School. As Glendora residents increased in number, so did the demand for new schools. In 1916, this first school was razed; today, there are nine K–12 schools in Glendora.

Citrus Union High School was founded on July 14, 1891, and an influx of students over the years outgrew the original high school in Azusa. On June 15, 1922, members of the Masonic Lodge laid the cornerstone of a new Citrus Union High School, built in a beautiful Spanish Renaissance style on Foothill Boulevard and Ben Lomond Avenue. This new facility was used until the late 1950s and early 1960s, when all of the original buildings were replaced with updated structures.

The Methodist church was organized in 1885 under Rev. H.J. Crall and was originally called the Alosta Charge of the Methodist Episcopal Denomination. Its first meetings were held at the LaFetra home on "Methodist Hill." A church was built in 1886 at Vista Bonita Avenue and Minnehaha Avenue (now Foothill Boulevard), moved to Bennett Avenue in 1893, and torn down in 1912. In its place, the Great White Temple was built by Horace Baker and officially dedicated on May 31, 1914.

Glendora Grace Episcopal Church began as a small church on Vista Bonita Avenue in November 1910 with only five families—Brunjes, Vickery, Banholt, Delancy, and Ward—and was the church that Rev. Henry Scott Rubel presided over starting in October 1935. In 1954, the original church building was moved to Mountain View Avenue, and in 1964, a new complex was consecrated, about when this photograph was taken.

When city hall was built in 1922, the property included the rest of the block to Carroll Avenue, deemed Glendora City Park. Shown here around 1960, the park is well established with original oak trees, covered picnic areas, and a variety of playground equipment. On February 26, 1972, the city park was replaced by a multi-million-dollar civic center including a public library, police station, and revamped city hall.

CHAPTER 6

GLENDORA, THE BEDROOM COMMUNITY

In a July 21, 1974, *Los Angeles Times* article, a deputy for the county's department of agriculture, W. David Dyer, claimed that nearly 800 acres of citrus groves were cleared in 1973, leaving only about 1,900 acres intact, mostly in the San Gabriel Valley. "There are no more citrus packing houses in Los Angeles County," said Dyer. They were mostly replaced with postwar housing tracts. This house, built around 1890, stood at the corner of Foothill Boulevard and Michigan Avenue.

The urban sprawl of the 1950s and 1960s saw a huge increase in houses being built and population growth in Glendora. In 10 years alone, between 1950 and 1960, Glendora grew from 3,988 people to 20,650. Shown here in 1952, Mayor Louis C. Wood and Chief of Police Oren Dan Fay Jr. visit a housing development site on Foothill Boulevard east of Minnesota Avenue. At the time, 400 new homes were under construction in Glendora.

The Haas house at the intersection of Pennsylvania and Sierra Madre Avenues is shown in 1910, with a Brush Runabout Model D 30 parked in front. The house was built by Jim Baker for Phillip and Hattie Haas in 1902, and in 1926, Frank and Leose Brown bought the ranch and took up citrus growing. When their land was subdivided around 1956, the surrounding streets, Brown Sage Drive and Leose Lane, were named in their honor.

James and Mary Odell came from Illinois in 1892 and built this house on Bennett and Washington Avenues on 15 acres of land where they planted apricots, peaches, and oranges. A year later, Frank and Elizabeth Odell moved to Glendora with their children. Frank bought the Kamphefner store on Glendora and Bennett Avenues, and in 1902 started the Glendora Bank with his father, James. It was the first brick building in town, and Frank was bank president until 1918.

After an arduous cross-country train trip, Issac and Lucetta Suydam settled in Glendora in 1887, building the town's first structure, a barn, in May of that year. They had two children, Keith and Nettie. This photograph was taken in 1909 and shows Keith and his wife, Lula, in front of their house on Vista Bonita Avenue, along with their daughter Hilda, who was born in the house, and Keith's mother, Lucetta. (Issac had died in 1908.) Keith was Glendora's first mail carrier.

From 1912 to 1923, Mary Miles Minter starred in 53 silent films. Her career ended abruptly when she was implicated in a murder in 1922. The year prior, at the height of her success, she purchased this home near the corner of Milton Drive and Barranca Avenue. The 45-foot-wide house was built in the 1910s and features two fireplaces made with stones from Dalton Canyon. Minter only lived in it for two years before retreating to Santa Monica.

Dr. August E. and Rosa Engelhardt moved into this house around 1895 on Colorado Avenue in the town of Alosta after the Engelhardt Brothers grocery and pharmacy burned down. It is a near duplicate of his brother Henry's house, as they are both believed to be Sears kit houses. Sometime after 1910, when this photograph was taken, the house was moved closer to the street. The Engelhardts lived here until 1924, when August died.

The Detwiler family arrived in the Glendora area around 1900 and built a house on then Alosta Avenue near Elwood Street. In 1902, they moved to this house on Whitcomb Avenue, east of Wabash Avenue and directly across the street from the Glendora Grammar School. The eight-person Detwiler family lived in this small house until 1948, when Mrs. Detwiler passed away. A porch, fireplace, some bedrooms, and living room were added during a remodel in the early 1950s.

Henry Engelhardt bought 40 acres in Glendora in 1882 and built a home at the southwest corner of Minnesota and Virginia Avenues nearly identical to that of his brother August. The top story was removed, and the house moved to Virginia and Cullen Avenues, where it now stands. Henry (with the beard) and his wife, Kate, are standing at center, and their three daughters, Anna, Clara, and May, are in front of them. Henry Kamphefner stands to the right.

The big "White House" of Glendora's founder, George Dexter Whitcomb, stood at the top of Glendora and Vista Bonita Avenues from 1886 to 1924. After Whitcomb died in 1914, the house was turned into a school for girls run by his daughter Leadora Dalzall. In 1924, the house burned down and was lost. The property became home to several military academies in the 1950s and 1960s and is now part of a church campus. All that remains of Whitcomb's house is a small rock retaining wall.

William H. Simmons was a real estate speculator who attended the Glendora land auction on April 1, 1887, and bought several lots in the new town. For reasons lost to history, Simmons did not return again to Glendora until 1889, just after the real estate bubble burst, when he built this house on Vermont Avenue north of Bennett Avenue. Though a dormer has been added to the second floor and the front door relocated, the scalloped siding on the gable remains.

On Sierra Madre Avenue, east of Minnesota Avenue, the Miller house is one of the oldest still-standing houses in Glendora. It was built in 1883 by Samuel J. Miller on 40 acres of land north of Sierra Madre. Taken in 1896, this photograph shows the Miller family (with Samuel at center in the rear) as well as Rev. F.A. Wight's family in the surrey. The original two-bedroom home was extensively renovated in the 1950s, but the façade remains the same. Note the eucalyptus tree.

This is the house of Charles A. Weaver, about whom little is known apart from the fact that he was a real estate agent, insurance salesman, and investor (he sold Empire automobiles in 1917). He was responsible for building, financing, or somehow supporting almost every large-scale building on Michigan Avenue from 1903 to 1910. Charles and his wife, Leila's, house is on the corner of Vermont and Meda Avenues. Attached to the Weaver house is a teahouse, and only the roofline can be seen from this angle.

In 1885, brothers Lawson and Milton H. LaFetra received Glendora land in lieu of payment for a bill from their grocery store in Los Angeles, which included around 500 acres. Eventually, five more LaFetra families joined what was called the LaFetra Colony northwest of Ben Lomond and Bennett Avenues. This house on Milton Drive was built by Clinton Norton in 1896 for Ira LaFetra's wife, Clara, and their children. Dick Davis bought the house in 1969 and lived in it until his death around 2017.

The house at Bennett and Vermont Avenues was built in 1904 by a wealthy easterner, Reuben Hamlin, who arrived in Glendora in the 1890s and became involved in local politics. When the first mayor of Glendora, J.S. Brubaker, died suddenly shortly after taking office in 1911, Hamlin was selected to take his place. Around this same time, Hamlin's wife, Helen, planted the bougainvillea vines on Bennett and Minnesota Avenues. Since 1982, the house has been the state headquarters of the Daughters of the American Revolution.

In October 1902, George Greiman purchased this house on the southwest corner of Vista Bonita and Ada Avenues from the Rice family. Shown here around 1907, the house has since gone through some internal improvements, but the exterior has not been significantly altered. George and his wife, Barbara, lived the rest of their lives in Glendora. Barbara passed away in 1924, while George lived on until 1938.

7

CHAPTER

Riding the Rails

For over 100 years, trains and trolleys were vital to everyday transportation. They could get you from one side of the country to the other, or from one side of the city to the other. Stations like Glendora's Pacific Electric (PE) station, built in 1908, dotted the tracks as the PE system spread across Southern California. The Pacific Electric system was operated at a loss, and when that loss became too great in 1950, the tracks and stations were abandoned.

The stately Victorian-style Glendora Station was the first impression for a lot of visitors who arrived to buy land or visit family. Completed in 1887, it was finished in panels of maple and redwood with artistic trim and Craftsman details. Shown here in 1940, a 1920-built 5000-class Baldwin locomotive is arriving. The station was replaced in 1947 and then torn down again in the 1960s. Townhomes now line the tracks, and a new Metro light rail station will soon be built.

On December 12, 1906, J.A. Jones, editor of the *Glendora Gleaner*, remarked, "It seems mighty nice to look out through the window from our desks and see the holes being dug . . . and the trolley wire strung and ties and rails being laid for the Pacific Electric line." Glendora officially became part of the expanding Pacific Electric system in December 1907. Although the last Red Car rumbled out of Glendora on September 30, 1951, the right-of-way could still be seen into the 1970s.

On the morning of March 15, 1911, these two 4-6-0-configured locomotives collided just east of the South Hills. The telegraph operator at the San Dimas station, Hayden Jones, neglected to hold the 481 train at a San Dimas siding so the 498 could pass. The result was a head-on collision, and fortunately, nobody was seriously hurt. Both engines were repaired; the 498 was scrapped in 1929, and the 481 stayed in service until 1938.

Accidents along the Pacific Electric lines were relatively common. What caused this Red Car to decouple and topple off the tracks is unknown, but it happened on the Glendora-Monrovia Line just west of Ben Lomond (now Barranca) Avenue around 1946 and drew quite a crowd. Note the LaFetra passenger shelter in the distance. By 1945, this line was servicing 2.7 million passengers a year. Today, work on the tracks at Barranca continues as new neighborhoods rise all around.

Designed by Edgar T. Wheeler and built in 1907, the Puente Largo Bridge was one of the most impressive structures on the entire Pacific Electric system. A reinforced concrete structure 1,019 feet long, it had 18 spans, each 57 feet long, 12 feet thick, and 30 feet wide. Each arch had a height of 19 feet. The bridge is now part of a bike and pedestrian path that links to the Lario Trail system in Duarte.

There were three Pacific Electric stations along the Monrovia-Glendora Line. Each was a wood-frame building containing a ticket office and waiting room, agent's office, and freight room. They were constructed for about $5,000 each. Built in 1907, Red Car 999 is seen here loading passengers around 1948, two years before meeting its doom at the scrapyard. The station lasted until about 1960, and in its place today is a relaxing park-like plaza on a bustling downtown Glendora Avenue.